Exploring Our Oceans

Written by Joanne Barkan

Table of Contents

Introduction 2
The Ocean Floor 6
Ocean Life 10

Introduction

From outer space, Earth looks like a blue ball. Can you guess why our planet looks so blue? It's because oceans cover most of Earth's surface. All this water comes from just five oceans: the Atlantic, Pacific, Indian, Arctic, and Southern Oceans.

These five oceans are all connected to one another and are actually part of one huge body of water called the "world ocean." This doesn't mean the oceans are exactly alike. In fact, they are different in many ways, including size.

How the Oceans Evolved

Millions of years ago, Earth had only one huge land mass, or continent. Over time, this land mass split apart into seven large continents. As these continents moved and changed, they formed the five oceans.

The Arctic is the smallest ocean. The Pacific is the giant of the group. It is almost large enough to hold the Atlantic, Indian, and Arctic Oceans within its borders. Let's explore and find out more about the oceans.

The Ocean Floor

Imagine exploring the bottom of the ocean. If you start at the shore, you'll find a flat, sandy floor that's close to the surface of the water. Here the sun warms the water, and the water is clear enough to see through.

As you go deeper, the ocean floor slopes further and further below the surface. At about 3,300 feet down, the sunlight disappears completely and the water turns icy cold and inky black.

Moving water in the ocean picks up salt from rocks—this helps make the water salty.

The surface layer of the ocean is full of sea life.

7

Down in the icy darkness, the ocean floor changes, too. You'll find miles of flat plains, wide valleys, and deep canyons, just like you see on land. Some of the tallest mountains in the world rise out of the ocean floor. And some of the deepest canyons, called **trenches**, drop thousands of feet below the waves.

There are also erupting underwater volcanoes that blast hot lava into the cold, dark water. Over many years, layers of this cooled lava build up and rise above the ocean's surface to become volcanic islands, such as the Hawaiian Islands.

Coral reefs often surround volcanic mountains.

Continental islands are pieces of land that were once attached to a continent.

Active volcanoes in Hawaii keep changing the shoreline's landscape.

The Deepest Trench

The Mariana Trench is the deepest spot in the ocean. It drops almost seven miles into the Pacific Ocean—that's more than six times deeper than the Grand Canyon. The Mariana Trench is also the lowest point on Earth's surface!

Ocean Life

The ocean is full of sea life. More than a million different kinds of plants and animals live in the ocean. Most of them need sunlight to survive, so they live in the upper layer called the Sunlit Zone.

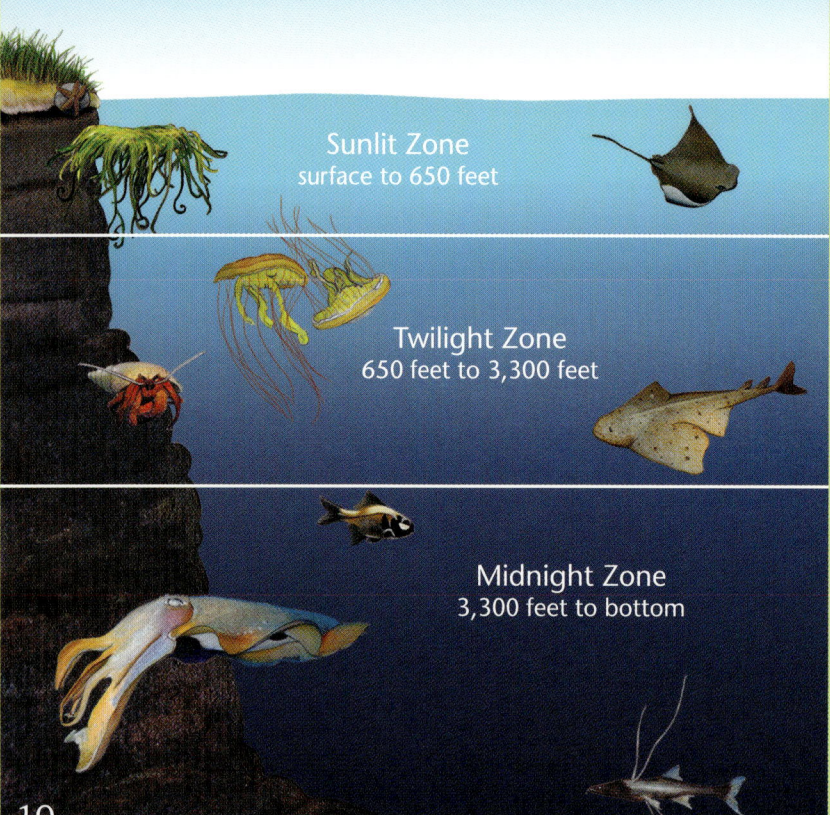

The Ocean Zones

Sunlit Zone
surface to 650 feet

Twilight Zone
650 feet to 3,300 feet

Midnight Zone
3,300 feet to bottom

Many seahorses make their homes in seaweed.

10

Seaweed-covered rocks at low tide

Bubble air sacs may help seaweed float toward the sunlight.

A lot of the plant life that you can see in the ocean is called seaweed. Besides seaweed, the other important ocean plant is phytoplankton. This plant is so tiny you need a microscope to see it! Seaweed and plankton are important sources of food for many ocean, or **marine**, animals.

Marine animals are all different sizes, shapes, and colors. They have many ways of surviving in the ocean. Saltwater fish and some **mollusks**, such as squid, spend their whole lives in the water. They breathe through **gills** and never have to come up for air.

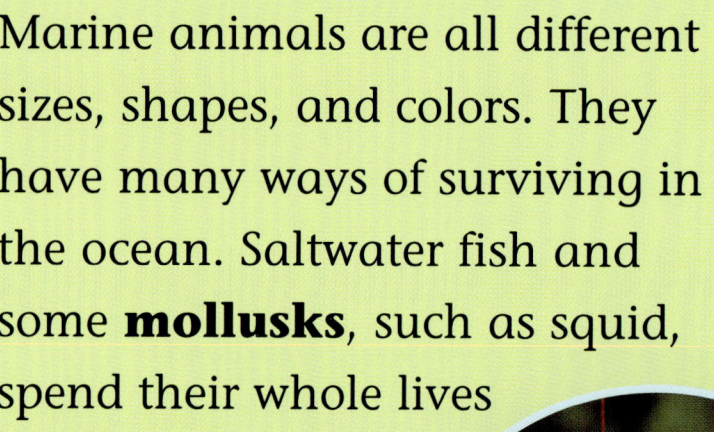

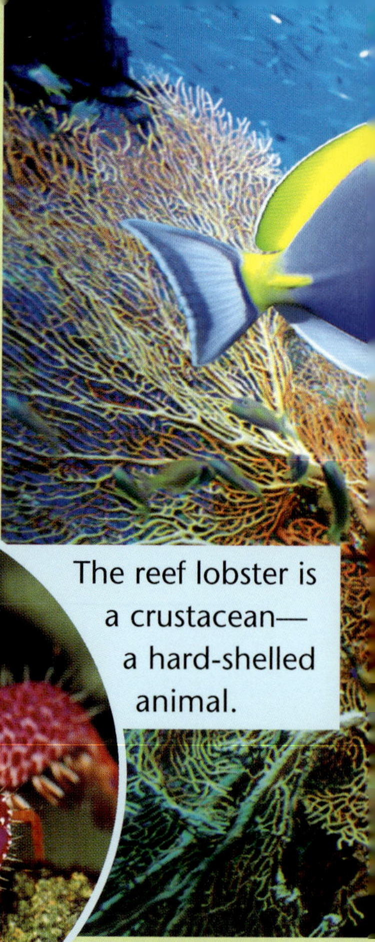

The reef lobster is a crustacean—a hard-shelled animal.

Marlin

Manta ray

Coral reefs are made from thousands of skeletons left by tiny animals called coral polyps.

Sea lion

Other animals also live in the water, but they must come up for air. Whales and porpoises come to the surface to breathe. They are **mammals** and breathe air through lungs. Sea lions and seals also breathe air, but these mammals live in the water and on the land.

Other animals live so far below the surface layer that they never come to the surface or see the sunlight. They have adapted to life thousands of feet below the surface. Many of these animals can glow in the dark, and others have long, flat bodies that help them glide along the bottom.

A scorpionfish can camouflage itself to hide from its prey.

The giant spider crab can measure 13 feet long when it stretches its legs—that's as long as a car!

Squid

Jellyfish

Anglerfish

Lots of them live in one of the most extreme underwater environments, the hot sea **vents**. About thirty years ago, scientists discovered that here in complete darkness, many animals such as crabs, tubeworms, and octopuses were able to live without plants or sunlight. They survive by feeding off the **bacteria** clumps in the boiling hot water that pours out of the vents.

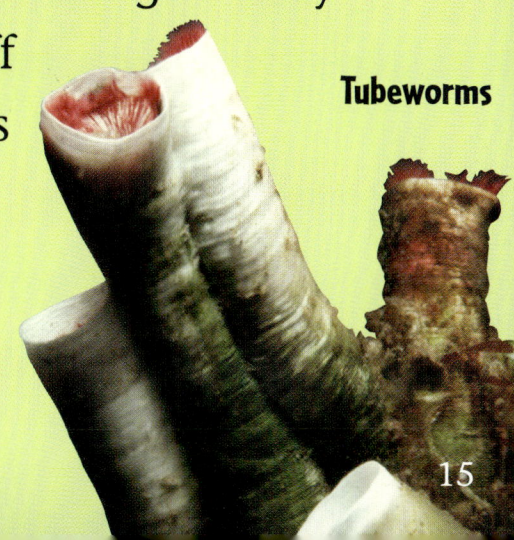
Tubeworms

15

Scientists have only explored a tiny part of Earth's ocean. They keep discovering new species of sea life all of the time. Have you ever discovered something new in the ocean?